CW00531537

A GUIDE

TO THE

1999

TOTAL ECLIPSE

OF THE

SUN

Steve Bell

H.M. Nautical Almanac Office
Royal Greenwich Observatory

LONDON: HMSO

© Copyright Particle Physics and Astronomy Research Council 1996
Applications for reproduction should be made to
HMSO, The Copyright Unit, St Clements House, 2-16 Colegate, Norwich NR3 1BQ

ISBN 0 11 702080 X

The photograph on the front cover of this booklet shows the inner corona of the Sun during the total eclipse of 18th February 1980 as observed from Kenya. Prominences can also be seen as red dots in the pearly white inner corona of the sun. © Michael Maunder.

DISCLAIMER OF LIABILITY

Neither PPARC, the RGO, its employees nor anyone else involved in the creation, production, or official distribution of this product, shall be liable for any direct, indirect, consequential, or incidental damages arising out of the use, the results of use, or inability to use this product even if the Royal Greenwich Observatory has been advised of the possibility of such damages or claim.

A Guide to the 1999 Total Eclipse of the Sun

CONTENTS

1	— The purpose of this booklet	5
2	— What is an eclipse?	5
3	— Predicting eclipses	6
4	— A brief anatomy of the Sun	8
5	— The last total eclipse visible in the UK	8
6	— The August 1999 total eclipse	8
7	— What do you see during a total eclipse of the Sun?	10
8	— Using the booklet	12
9	— Local circumstances of the partial eclipse	12
10	— Local circumstances of the total eclipse	16
11	— Observing the eclipse	20
12	— Weather in the path of totality	22
13	— Future eclipses of the Sun in the UK	22
14	— A quick summary	23
15	— The aluminized mylar eclipse viewer	24
16	— Useful references	24

ACKNOWLEDGMENTS

The coastal outlines used in this booklet have been created using Autoroute Plus, a product of the Microsoft Corporation which is based on mapping provided by Ordnance Survey with permission of the Controller of Her Majesty's Stationery Office. Weather information has been provided in part by the National Meteorological Library and Archive.

I would like to thank my colleagues in H.M. Nautical Almanac Office and the Royal Greenwich Observatory for their assistance in the production of this booklet and to all those who have made suggestions leading to the improvement of the content of the booklet.

Steve Bell
H.M. Nautical Almanac Office
1996 August

PUBLICATIONS OF H.M. NAUTICAL ALMANAC OFFICE

The Astronomical Almanac contains ephemerides of the Sun, Moon, planets and their natural satellites, as well as data on eclipses and other astronomical phenomena.

Astronomical Phenomena contains data on the principal astronomical phenomena of the Sun, Moon and planets (including eclipses), the times of rising and setting of the Sun and Moon at latitudes between S 55° and N 66°, and calendarial data.

The Nautical Almanac contains ephemerides at an interval of one hour and auxiliary astronomical data for marine navigation.

The Air Almanac contains ephemerides at an interval of ten minutes and auxiliary astronomical data for air navigation.

Sight Reduction Tables for Air Navigation (AP3270), 3 volumes. Volume 1, selected stars for epoch 1995·0, containing the altitude to 1′ and true azimuth to 1° for the seven stars most suitable for navigation, for the complete range of latitudes and hour angles of Aries. Volumes 2 and 3 contain values of the altitude to 1′ and azimuth to 1° for integral degrees of declination from N 29° to S 29°, for the complete range of latitudes and for all hour angles at which the zenith distance is less than 95° providing for sights of the Sun, Moon and planets.

Planetary and Lunar Coordinates, 1984–2000 provides low-precision astronomical data for use in advance of the annual ephemerides and for other purposes. It contains heliocentric, geocentric, spherical and rectangular coordinates of the Sun, Moon and planets, eclipse data, and auxiliary data, such as orbital elements and precessional constants. (The next edition for 2001–2020 is being prepared).

All the above publications are prepared jointly by H.M. Nautical Almanac Office, Royal Greenwich Observatory, and the Nautical Almanac Office of the United States Naval Observatory, and are published jointly by HMSO and the United States Government Printing Office.

Sight Reduction Tables for Marine Navigation (NP 401), 6 volumes. This series is designed to effect all solutions of the navigational triangle, given two sides and the included angle to find the third side and an adjacent angle; the tables are arranged to facilitate rapid position finding and are intended for use with *The Nautical Almanac*. Explanatory material and auxiliary tables are included in all volumes.

The Star Almanac for Land Surveyors contains the Greenwich hour angle of Aries and the position of the Sun, tabulated for every six hours, and represented by monthly polynomial coefficients. Positions of all stars brighter than magnitude 4·0 are tabulated monthly to a precision of $0\overset{s}{.}1$ in right ascension and 1″ in declination. Coefficients for calculating the Greenwich Hour angle and declination (Dec) of all the stars published in *The Star Almanac for Land Surveyors*, accurate to about 1″, for 1997 are available on floppy disk. The polynomial coefficients for calculating R, Dec, E and semi-diameter of the Sun are also included. This publication is available from HMSO and from UNIPUB, 4611/F Assembly Drive, Lanham, MD 20706-4391, USA.

Compact Data for Navigation and Astronomy for 1996–2000 contains data, which are mainly in the form of polynomial coefficients, for use by navigators and astronomers to calculate the positions of the Sun, Moon, navigational planets and bright stars using a small programmable calculator or personal computer. A 3.5-inch disk for IBM PC and compatibles is included. It contains the astronomical tabular data in ASCII files and NAVPAC, an interactive software package based on the methods in the book. This publication is available from HMSO and from UNIPUB, 4611/F Assembly Drive, Lanham, MD 20706-4391, USA.

Further details about the publications and services provided by H.M. Nautical Almanac Office may be found via Internet on the World Wide Web (WWW). The home page address is http://www.ast.cam.ac.uk/~nao/.

THE 1999 TOTAL ECLIPSE OF THE SUN

1 — The purpose of this booklet

On the morning of Wednesday 11th August 1999, a total eclipse of the Sun will be visible from the south-western part of the UK mainland, the Scilly Isles and the Channel Islands lasting up to 2^m 06^s. The path of totality starts at sunrise to the south of Nova Scotia and crosses the Cherbourg peninsula, northern France, the southern tip of Belgium, Luxembourg, southern Germany, Austria, Hungary, the north-eastern tip of Yugoslavia, Romania, the north-eastern part of Bulgaria and the Black Sea. It then crosses over central Turkey, the north-eastern part of Iraq, Iran, southern Pakistan, central India and ends at sunset over the Bay of Bengal. A partial eclipse will be seen by the north-eastern part of North America, Greenland, Iceland, the rest of the UK, the Irish Republic, the remainder of Europe, North Africa, the Middle East and much of Asia as far east as Thailand and central China.

This guide has been produced with the aim of helping people to get the most out of this spectacle. Although the eclipse has the potential of being seen by the largest number of people in the history of eclipse watching, this guide will concentrate on the event as it will be seen from the British Isles. Assuming the weather conditions are good, this will be the only opportunity to see a total eclipse of the Sun for nearly a century from this part of the world.

Sections on how eclipses occur and how they can be predicted are supplied. A short description of the Sun is given to explain how a total eclipse allows observers to see parts of the atmosphere of the Sun that are normally invisible. A description of the phenomena that occur during a total eclipse of the Sun is provided as well as several examples demonstrating the use of the tables and figures in this guide. The local circumstances of the eclipse are supplied for many places across the UK, the Irish Republic and the Channel Islands. Information on how to observe the eclipse safely is also provided.

2 — What is an eclipse?

As the Moon orbits the Earth it is said to be in opposition when the Sun and Moon are diametrically opposite each other in the sky. This occurs when the Moon is full. When the Sun and Moon lie in the same direction they are said to be in conjunction and the Moon is new. In the course of a year, there are occasions when the Sun, Earth and Moon lie in a straight line. When this happens one body will deprive the other of sunlight to some degree causing an eclipse. If the orbits of the Earth and the Moon were in the same plane, lunar eclipses would occur every full Moon and solar eclipses every new Moon. As the orbital plane of the Moon is inclined at 5° to that of the Earth, the frequency with which these alignments take place is much reduced. For an eclipse to occur the Moon has to be close to one of two points where the orbital planes of the Earth and Moon intersect in addition to being either new or full. If the new Moon is at either of these intersections, a solar eclipse will take place. If the Moon is full, a lunar eclipse will occur.

The Sun is 400 times the diameter of the Moon but it is also 400 times further away from the Earth. This fortuitous coincidence means that on average the two objects appear to have similar angular sizes in the sky, a unique occurrence in the solar system. If the orbits of the Earth and Moon were circular, solar eclipses would always be total as the relative distances between the Earth and the Sun and the Earth and the Moon would be constant. In reality, the orbits of the Earth and Moon are ellipses or elongated circles. As a result, the apparent angular diameter of the Sun varies by about 2% whereas that of the Moon varies by about 8%. If an alignment takes place when the Moon is further away than its average distance from the Earth, its apparent diameter is not sufficient to obscure the Sun completely. This causes the central portion of the Sun to be eclipsed leaving a bright ring of sunlight around the Moon. This type of eclipse is known as an annular eclipse.

If the alignment occurs when the Moon is nearer the Earth than its average distance, a total eclipse takes place. Sometimes the umbral shadow of the Moon may reach the Earth only when it lies very near the line of centres of the Sun and Earth. At either end of the eclipse track, the shadow is not long enough to reach the Earth. In these circumstances we see an eclipse that starts as an annular one, becomes total and then reverts to being annular once more. This hybrid eclipse is known as an annular-total eclipse and occurs relatively rarely.

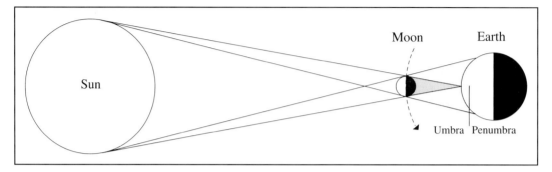

Figure 1: The configuration of the Sun, Moon and Earth during a total eclipse of the Sun. This is a schematic diagram and is not drawn to scale. The umbral shadow is the darker grey area between the Moon and the Earth and the penumbral shadow is the light grey area. A total eclipse will be seen in the umbral shadow whereas a partial eclipse will be seen in the penumbral shadow.

The configuration for a total eclipse of the Sun is shown in Figure 1. The Sun is not a point source of light and produces umbral and penumbral shadows when eclipsed by the Moon. In an umbral shadow, none of the light from the Sun can be observed from a point within the shadow. A penumbral shadow is formed when light from the Sun is not completely cut off by the Moon. If the umbral shadow reaches the Earth then a very small portion of the Earth's surface will see a total eclipse. As all the bodies are in motion, the umbral shadow travels rapidly across the face of the Earth from west to east. The movement of the shadow is known as the path of totality. Those areas in the penumbral shadow will see a partial eclipse. In the case of an annular eclipse, the umbral shadow just fails to reach the surface of the Earth. If the alignment of the Sun, Moon and Earth is not sufficiently precise, then a partial eclipse will be seen by some fraction of the Earth's surface.

In the case of lunar eclipses, any observer on the Earth can see an eclipse of the Moon as long as the Moon is above the horizon. The shadow cast by the Earth is much larger than the Moon itself and we can observe either penumbral or umbral lunar eclipses. Of these two types of eclipse, penumbral eclipses of the Moon are generally more difficult to detect as the resulting dimming of light tends to be quite small. The majority of observed lunar eclipses, both partial and total, tend to be umbral eclipses. Many people will have seen the copper red appearance of the Moon during a total lunar eclipse. Comparing the durations of solar and lunar eclipses, we see that a total eclipse of the Sun can be as long as $7^m 31^s$ and an annular eclipse can last $12^m 30^s$, but a total lunar eclipse can last 100 minutes or more.

3 — Predicting eclipses

How can eclipses be predicted? The general circumstances required for a solar eclipse to take place is described in Section 2. The new Moon must lie near the points of intersection between its orbit and that of the Earth. These points are known as the nodes and are important in characterizing the orientation of the orbit. These nodes are not stationary but move in the opposite direction to the motion of the Moon at little over $0°05$ per day. Consequently, the Moon takes 27·21 days to move from one node back to the same node again. This interval is known as a Draconic month. For the Moon to pass from one new Moon to the next takes 29·53 days, an interval known as a lunation. If the new Moon was at one of its nodes, it would take 242 Draconic months or 223 lunations to get back to the same configuration. In this interval of a little over 18 years and 10 days, the Sun returns to the same node of the Moon's orbit nineteen times. Consequently, if an eclipse took place at the beginning of this cycle, another eclipse would occur some 18·03 years later. This cycle, known as the Saros, was known to the Babylonians more than 2500 years ago through their observations of lunar eclipses.

However, the Babylonians noticed that eclipses of similar types occurred at intervals of 18 years and 10 or 11 days later. For one annular eclipse to follow another on an 18 year time scale, the Moon must be further away than its average distance from the Earth on both occasions. Similarly for two total eclipses to follow each other, the Moon must be closer than its average distance from the Earth on both occasions. Hence, another cycle must play a part in the Saros. We know that for the Moon to go once round its elliptical orbit takes 27·55 days, a period known as an anomalistic month. During the Saros cycle, the Moon can complete 239 orbits and can therefore return to a similar position in its orbit and hence a similar distance from the Earth in a little over 18 years. Consequently, all three cycles mesh together to permit the recurrence of similar types of eclipses at 18 year intervals.

Each Saros is composed of between 70 and 85 eclipses and has a lifetime of about 1300 years. At any one time 42 Saros cycles are running simultaneously. The Saros starts with partial eclipses in the polar regions of the Earth, builds up to annular or total eclipses in the equatorial regions and ends in partial eclipses at the opposite pole. The Saros is not perfect because the underlying lunar cycles are not totally synchronized. Consecutive eclipses in a Saros do not occur at the same geographical location, they are shifted westwards by about 120° in longitude. The match between eclipse circumstances is better over three Saros cycles although the eclipse tracks are now shifted either north or south. The 54 year interval is called the Triple Saros.

Modern techniques for calculating eclipses use high precision numerical calculations such as those produced by NASA's Jet Propulsion Laboratory to determine the positions of the Earth and Moon relative to the Sun. These precise positions are known as ephemerides. Using techniques developed by F.W. Bessel, a nineteenth century German mathematician, it is possible to characterize the geometric position of the shadow of the Moon relative to the Earth. Solar eclipse maps can be generated from these calculations such as those reproduced in *The Astronomical Almanac*. These maps show the central path if any, the region of visibility of the eclipse and the timings of specific phases of the eclipse. The appearance of an eclipse from a specific location or "local circumstances" can also be determined.

Universal time (UT) is the basis of civil time-keeping and is defined by the daily motion of the stars due to the rotation of the Earth. Leap seconds are added about once a year at appropriate times to keep UT in step with the changes in the rotation of the Earth. Hence UT is not a uniform time scale. The ephemerides that are used to predict eclipses, however, are based on a uniform time scale called dynamical time. To obtain UT from dynamical time, a correction ΔT, (pronounced delta tee) has to be added to dynamical time. Unfortunately, ΔT cannot be predicted with any certainty, instead it has to be determined from astronomical observations made near to the time required. In this booklet we are adopting a value of $\Delta T = 65.5$ seconds for the total eclipse of 11th August, 1999. In the context of this booklet, Greenwich mean time (GMT) is the same as UT. However, during the summer months British summer time (BST) is in force which is one hour ahead of GMT. As the eclipse takes place during August, **all times quoted in this booklet are given in BST**.

What conditions are necessary for a good solar eclipse? The best opportunity to get a total eclipse of the Sun is when the Moon is closest to the Earth and its angular size is greatest. The apparent diameter of the Sun should also be as small as possible. This occurs when the distance from the Earth to the Sun is greatest, a point in the Earth's orbit called aphelion which occurs in early July. To prolong the eclipse as much as possible, it should take place in the equatorial regions of the Earth where the surface velocity reaches a maximum value of 1700 kilometres/hour and cancels out some of the motion of the shadow of the Moon which moves at approximately 3400 kilometres/hour. The Moon is also slightly closer to us when it is at the zenith or directly overhead. This can extend totality by a few seconds. At best, the width of the path of totality can reach 269 kilometres and the duration of totality could last $7^m\ 31^s$. In the case of an annular eclipse, the track can be as wide as 370 kilometres with eclipses lasting up to $12^m\ 30^s$.

4 — A brief anatomy of the Sun

The Sun is a normal star and is special only in that it is so close to us. In many respects, this makes it easier to study than other stars because the closest of them is some 250 000 times further away. A total eclipse of the Sun gives observers on the Earth an opportunity to study the extended atmosphere of the Sun which is invisible under normal circumstances. In order to understand some of the phenomena we see at an eclipse, a brief description of the structure of the Sun is needed.

The Sun's energy is generated in a very dense core about the size of Jupiter where the temperature reaches 15 million degrees Celsius. Surrounding this core is a deep layer of hydrogen which provides the fuel for the nuclear reactions taking place in the core. The flow of energy from the core passes through progressively less dense layers of the Sun until it reaches a region called the convective zone where the flow becomes turbulent. This region is the source of the solar magnetic field. Above this layer is the visible face of the Sun called the photosphere. It is only 500 kilometres thick (0·1% of the solar radius) and is the home of such familiar features as sunspots. We cannot see deeper into the Sun than this layer as it is completely opaque. This is the reason why the Sun appears to have a sharply defined edge. Light from the photosphere must then pass through two more regions to reach us, the chromosphere or "colour-sphere" and the corona.

On average, the chromosphere is about 4000 kilometres thick. The name is derived from its pinkish colour caused by an atomic transition of hydrogen. Beyond this layer is the corona which extends millions of kilometres into space. It is a region of extremely hot gas whose density is so low that one gramme of material occupies more than one cubic kilometre. Both the chromosphere and the corona can only be observed by the naked eye during eclipses unless special equipment is used. Disturbances in the chromosphere called spicules can reach 15 000 kilometres into the corona but these substantial features are overshadowed by the more spectacular prominences. These arches of glowing gas can reach 250 000 kilometres into the corona and can sometimes be seen during totality. When the Sun is at its most active, the corona is a complex structure exhibiting many streamers which can extend many solar radii into space. Early observations of the shape of the corona pointed to the existence of magnetic fields in the Sun long before they were measured. The pattern of the streamers in the corona bear a strong resemblance to the configuration that iron filings adopt in the presence of a bar magnet. At the minimum of the solar cycle, much of the activity is confined to the solar equator.

5 — The last total eclipse visible in the UK

With the exception of the total eclipse of 30th June 1954 which was visible only in the northernmost part of the Shetland Islands, the last total eclipse of the Sun visible in the United Kingdom took place nearly seventy years ago. On 29th June 1927, not long after sunrise, the path of totality crossed the coast of Cardigan Bay to the west of Portmadog, passed to the east of Colwyn Bay and out over the Irish Sea. It crossed the coastline again at Southport and moved north east over Settle, Richmond, Darlington and West Hartlepool and then out over the North Sea. Totality lasted a little less than 25 seconds and was seen from a strip of land less than 50 kilometres wide. For the rest of the UK, the eclipse was seen as a partial one.

6 — The August 1999 total eclipse

Fortunately, the total eclipse of 11th August 1999 takes place during late morning when the Sun will be relatively high in the sky. It will be visible from the Scilly Isles, most of Cornwall, western Devon and Alderney in the Channel Islands. The total eclipse will be seen from a strip of land a little over 100 kilometres wide and totality will last a maximum of $2^m 06^s$ in the British Isles.

In a global context, the path of totality starts at sunrise approximately 400 kilometres south of Halifax, Nova Scotia and crosses the Atlantic where its first landfall is the Scilly Isles followed closely by south-west England. After passing over Alderney in the Channel Islands, it crosses the Cherbourg peninsula, passing over northern France, the southern tip of Belgium, Luxembourg, southern Germany, Austria, Hungary, the north-eastern tip of Yugoslavia, Romania, the north-eastern part of Bulgaria and out over the Black Sea. The path of totality then crosses over central Turkey, the north-eastern part of Iraq, Iran, southern Pakistan, central India and ends at sunset over the Bay of Bengal approximately 500 kilometres east of the Indian city of Srikakulam. Maximum eclipse takes place over Romania at $12^h 03^m$ BST where totality reaches a maximum duration of $2^m 27^s$. The path of totality is shown in Figure 2. Two capital cities will see totality, namely Luxembourg and Bucharest while Paris, Vienna and Budapest will all see more than 99% of the Sun disappear. The German cities of Stuttgart and Munich are also on the path of totality. A partial eclipse of the Sun will be seen in the north-eastern part of North America, Greenland, Iceland, Europe, North Africa, the Middle East and much of Asia as far east as Thailand and central China.

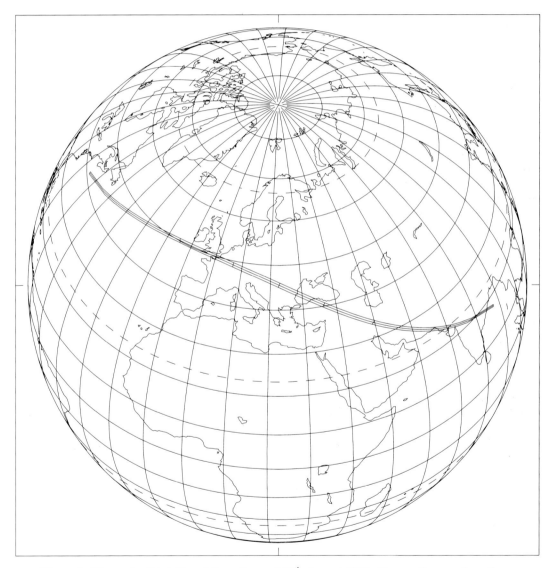

Figure 2: The path of totality of the eclipse of 11th August 1999. The northern and southern limits of totality as well as the central line of the eclipse are shown passing through the central section of the diagram. The vertical bar to the south west of Ireland represents the point of maximum eclipse at $11^h 00^m$ BST. The vertical bars are spaced at 30 minute intervals.

In a historical context, this eclipse is the 21st in a total of 77 eclipses in Saros 145. The series began on 4th January 1639 as a marginal partial eclipse in the North Polar regions. It will end on 17th April 3009 in a similar fashion in the South Polar regions. As Saros 145 matures, the duration of totality will increase and eclipse tracks will move gradually towards the equator.

7 — What do you see during a total eclipse of the Sun?

The beginning of a solar eclipse is known as "first contact". This is the moment when the Moon begins to obscure the Sun's photosphere, giving the appearance of a "bite" taken from the edge of the very bright solar disk. The "partial phase" follows where progressively more and more of the Sun's photosphere is obscured by the Moon. Shortly before totality, the Moon's shadow can be seen approaching from the western horizon giving the impression of an approaching storm. The partial phase lasts about an hour or so until "second contact" when the Sun is hidden by the Moon leaving the last remnants of the Sun's photosphere as a thin silvery ring around the limb of the Moon.

During the partial phases of the eclipse, you may see many hundreds of images of the crescent Sun on the ground beneath trees. This phenomenon is caused by the gaps in the foliage acting as pinhole cameras focusing the crescent image of the Sun. In the few minutes before second contact, the sky darkens noticeably and both flora and fauna react to the increasing darkness. Some flowers may close up, animals may behave as they would at nightfall and birds may go to roost. As the amount of light diminishes more rapidly, the landscape can take on a metallic grey hue. The temperature may also drop by a couple of degrees. An elusive phenomenon known as "shadow bands" may be observed which give the appearance of parallel light and dark bands moving rapidly across the ground. These ripples are caused by irregular refraction of the crescent shaped image of the Sun and may appear again in the few minutes after "third contact" or the end of totality.

Just before totality, the bright photospheric ring around the Sun breaks up into discrete blobs of light known as Baily's Beads. This effect is caused by the final flashes of sunlight shining through some of the gaps between the lunar mountain ranges on the eastern limb of the Moon. Within a few seconds all the "beads" disappear bar one. Nearly all of the bright photosphere is gone and the Sun's corona becomes visible as a pearly white ring, sometimes irregular in shape, around the Sun. The final bright spot of the photosphere in conjunction with the ring formed by the inner corona give rise to the so-called "diamond ring effect". Within seconds this feature disappears and the corona comes into full view. Totality has now begun.

During totality, the entire horizon may appear orange or maroon resembling the colours of the sky after sunset. The colour of the sky in the direction from which the shadow of the Moon is approaching is usually a purple colour. Bright stars and planets can be seen as the sky grows dark. The Sun's corona can now be seen clearly, extending radially away from the Sun in all directions. It is a pearly white colour and may extend several solar radii from the Sun depending on how active the Sun is at the time of the eclipse. At the minimum of solar activity, the Sun's magnetic field binds the coronal gas into streamers. These streamers are wide at their base close to the Sun and curl up to a point. At the poles of the Sun, the streamers take on the appearance of thin streams of gas, like iron filings following the magnetic field of a bar magnet. The more active the Sun, the greater the number of streamers from a wider range of latitudes. When the Sun is at its most active, the corona may appear as a ring-like feature. As the Sun is expected to be close to maximum activity in 1999, we can look forward to an active corona during this particular total eclipse. You may be able to see the chromosphere of the Sun as a pinkish ring around the edge of the Moon's disk. Sometimes, when the Sun is particularly active you might be able to see signs of prominences, pinkish arcs of gas within the inner regions of the corona and the chromosphere.

At "third contact", the total phase of the eclipse is over. What was seen in the moments before totality now occurs in reverse time order. A second "diamond ring" may appear followed by another display of Baily's Beads on the western limb of the Moon. The Moon's shadow can be seen heading towards the eastern horizon shortly after totality is over. The Sun's photosphere gradually becomes more and more dominant and the eclipse finishes when the silhouette of the Moon disappears at fourth contact. Examples of what you might see during the eclipse are shown in Plates 1 – 4.

Figure 3 shows the configuration of the Sun and Moon at different times during the eclipse and also the points on the Moon's limb where Baily's Beads may occur. The first panel shows the relative positions of the Sun and Moon at first contact. The Moon is shown as a black disk simply to illustrate its position. In reality, we can only detect its position by the way it obscures the Sun's disk. Forty minutes after first contact, the Sun will appear as a crescent shape as indicated in the second panel.

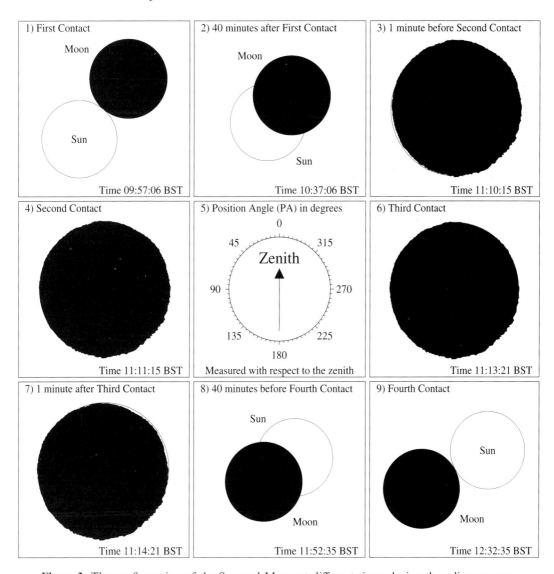

Figure 3: The configuration of the Sun and Moon at different times during the eclipse as seen from Falmouth. The features on the lunar limb and the size of the Moon's disk have been exaggerated for the panels at or near the times of second and third contact for the reasons given in Section 7. This also illustrates how features on the lunar limb will give rise to Baily's Beads just before second contact and just after third contact. The central panel shows how position angle (PA) is defined in this booklet. Further information on position angles can be found in Section 9. Although this diagram shows the local circumstances for Falmouth, the overall appearance of the eclipse should be similar at any point on the path of totality over the UK and the Channel Islands although the times will be different.

In the third panel, we have increased the sizes of the Sun and Moon relative to the two previous panels to illustrate the effect of the topography of the Moon on the lunar limb at an instant one minute before second contact. The profile of the Moon is based on measurements by C.R. Watts derived from a photographic atlas of the Moon. The features on the lunar limb have been exaggerated to show the lunar valleys and mountains more clearly. Baily's Beads may be visible where there is a sizeable gap between the solid outline of the Moon and the photospheric disk of the Sun.

At second contact, the Moon completely obscures the Sun. This is shown in the fourth panel on the same scale as the previous panel. The central panel illustrates how position angle is defined in this booklet and its relationship to the zenith. A more thorough discussion of both of these terms can be found in Section 9. The sixth panel shows the situation at third contact and the next panel shows the reappearance of the Sun's disk one minute after third contact. The last two panels show the relative positions of the Sun and Moon forty minutes before fourth contact and at fourth contact on the same scale as the first two panels. Although this diagram shows the local circumstances for Falmouth, the overall appearance of the eclipse should be similar at any point on the path of totality over the UK and the Channel Islands although the times will be different. Local circumstances will be discussed more fully in Sections 9 and 10.

For those people who can observe the total eclipse but are located close to the edge of the path of totality, there is some compensation for the brevity of the total phase of the eclipse. The duration of the display of Baily's Beads increases as you move closer to the edge of the path of totality. The southern edge of the path will provide a better display of beads because the lunar landscape is more rugged near the lunar south pole.

8 — Using the booklet

The next two sections of this booklet describe how the eclipse will be seen from different parts of the UK, the Irish Republic and the Channel Islands. The timing and appearance of an eclipse from a particular place are known as "local circumstances". These circumstances will be different for everyone. As the path of the eclipse moves eastwards at supersonic speeds so observers in Cork will see maximum eclipse more than 13 minutes before those in Norwich. The majority of observers in the British Isles will see a partial eclipse and the Section 9 includes a discussion of the eclipse as they will see it. Those observers lucky enough to be in the path of totality can find a description of their view of the eclipse in Section 10.

Perhaps the two most important pieces of information about the eclipse are shown in Figure 4. The lines drawn almost vertically join together points having the same times of maximum eclipse, the moment when the area of the Sun's disk obscured by the Moon is greatest. The nearly horizontal lines join together places seeing the same degree of obscuration at maximum eclipse. Symbols showing the relative positions of the Sun and Moon at maximum eclipse for the relevant degrees of obscuration are given at the right hand side of the map. At the bottom of the diagram the northern and southern limits of the path of totality have been plotted as well as the central line of the eclipse passing over south-west England. The Channel Islands have not been shown in Figure 4. However, the southern limit of the path of totality and the times of maximum eclipse relevant to the Channel Islands have been plotted in the upper left hand corner of Figure 5.

To illustrate the purpose of Figure 4, let us use a simple example. It can be seen that Lincoln, Aberdeen and Kirkwall will all experience the greatest obscuration of the Sun at approximately $11^h 20^m$ BST. However, Lincoln will see an obscuration of 90·7%, Aberdeen will see 77·6% and Kirkwall will see only 71·9% of the Sun disappear. To put these figures into perspective, an obscuration of 85% will cause the ambient light level to drop noticeably. As a result most of UK and the Irish Republic will see a measurable darkening of the sky. However, in the north of Scotland, the effect will be much smaller.

9 — Local circumstances of the partial eclipse

Table 1 gives local circumstances for 110 cities and towns across the UK and the Irish Republic. These locations have been selected to provide a reasonable geographical distribution over the two countries. As the eclipse will be seen as a partial one from these places, times of first and fourth contact have been given. Second and third contact are only applicable to those locations seeing a phase of totality. In order to explain the meaning of the information given in Table 1, let us use Cambridge as an example.

The first column in the table refers to the location and the next four columns refer to first contact. The time of first contact or the beginning of the eclipse is $10^h 04^m 32^s$ BST. The position of the Sun in the sky is described by its azimuth and altitude. Azimuth (Az) is measured clockwise from true north through east, south, west and back to north.

Plate 1: If there are trees near the location from which you are observing the total eclipse of 11th August 1999, many images of the crescent Sun may be visible in the shadows beneath those trees during the partial phase of the eclipse. This phenomenon is caused by gaps in the foliage acting as pinhole cameras which focus the image of the Sun. In this 10th May 1994 photograph of an annular eclipse taken in Arizona, many images of the eclipsed Sun can be seen.

© *Michael Maunder*

Plate 2: Photographs of solar eclipses need not simply feature the Sun. Capturing the effect that the total eclipse has on your surroundings can be just as rewarding. In this wide-angle photograph of the total eclipse of the Sun taken on the morning of 23rd October 1976 from Zanzibar, the eclipsed Sun can be seen against the inverted cone-shaped shadow of the Moon.

© *Michael Maunder*

Plate 3: One of the most beautiful sights to be seen during a total eclipse of the Sun is the so-called 'diamond ring effect'. This feature is caused by the last rays of sunlight passing between features on the lunar limb. This photograph was taken using a Celestron C90 1000 millimetre f/11 telescope at the moment of third contact during the eclipse of 3rd November 1994 in Chile. Prominences can also be seen as red dots in the pearly white inner corona of the Sun.

© Nick Quinn

Plate 4: Without special equipment, the Sun's corona can only be seen during a total eclipse of the Sun. This photograph was taken during the total phase of the eclipse of 11th July 1991 as seen from Baja California using a Celestron C90 1000 millimetre f/11 telescope. As the Sun was very close to the maximum of its activity cycle, the corona can also be seen as a broad feature surrounding the Sun exhibiting considerable structure. Streamer activity can also be seen very clearly.

© *Nick Quinn*

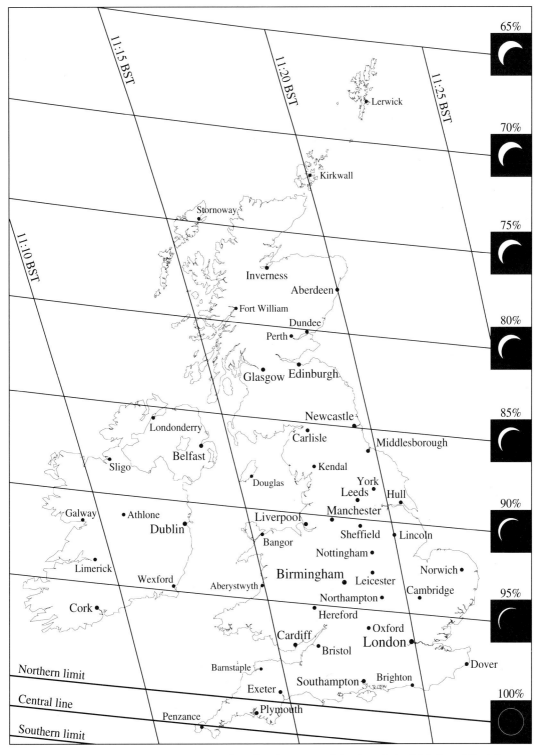

Figure 4: The circumstances of the eclipse for the UK mainland and the Irish Republic. The near vertical lines link sites having the same times of maximum eclipse whereas those running nearly horizontally link places experiencing the same degree of obscuration at maximum eclipse. The appearance of the Sun at maximum eclipse is shown for the relevant obscuration on the right hand side of the diagram. The ring shown in the 100% obscuration diagram is a simulation of the appearance of the corona. The northern and southern limits of the path of totality are indicated as well as the central line of the eclipse. *Coastal outlines © Crown copyright*

Table 1: Partial eclipse local circumstances

Place	First Contact				Maximum Eclipse					Fourth Contact			
	Time	Az	Alt	PA	Time	Az	Alt	Mag	Obs	Time	Az	Alt	PA
	h m s	°	°	°	h m s	°	°			h m s	°	°	°
Aberdeen	10 08 26	121	36	300	11 20 02	141	43	0·818	0·776	12 34 39	166	48	123
Aberystwyth	10 00 32	114	36	314	11 14 56	134	45	0·949	0·942	12 33 36	161	52	120
Athlone	09 58 20	110	33	312	11 10 47	129	42	0·924	0·912	11 27 46	154	50	125
Ayr	10 03 52	116	35	306	11 16 06	136	43	0·866	0·837	12 32 00	161	49	124
Bangor, Gwynedd	10 01 26	114	36	312	11 15 20	135	45	0·927	0·914	12 33 20	161	51	121
Barnstaple	09 59 04	113	36	317	11 14 12	133	46	0·986	0·987	12 33 54	161	53	119
Barrow-in-Furness	10 03 22	116	36	309	11 16 58	137	45	0·901	0·881	12 34 21	163	50	121
Bath	10 01 05	115	37	316	11 16 41	136	47	0·975	0·975	12 36 30	165	53	117
Belfast	10 01 35	114	34	308	11 14 01	133	43	0·891	0·869	12 30 29	158	49	124
Berwick-on-Tweed	10 06 37	120	36	304	11 19 22	140	44	0·855	0·823	12 35 23	166	49	122
Bideford	09 58 52	112	36	318	11 13 58	133	46	0·988	0·990	12 33 41	160	53	119
Birmingham	10 02 50	117	37	313	11 17 57	138	46	0·943	0·935	12 36 57	166	52	118
Blackburn	10 03 38	117	37	310	11 17 43	138	45	0·910	0·893	12 35 33	165	51	120
Blackpool	10 03 12	116	36	310	11 17 03	137	45	0·908	0·891	12 34 42	164	51	121
Bournemouth	10 00 53	115	38	318	11 17 01	136	48	0·993	0·995	12 37 26	165	54	116
Bradford	10 04 24	118	37	309	11 18 41	139	46	0·907	0·890	12 36 35	166	51	119
Brighton	10 02 52	117	39	317	11 19 33	139	49	0·986	0·988	12 40 11	169	54	114
Bristol	10 00 56	115	37	316	11 16 25	136	47	0·974	0·973	12 36 09	164	53	117
Bude	09 58 20	112	36	318	11 13 24	132	46	0·993	0·995	12 33 10	160	53	119
Cambridge	10 04 32	119	39	313	11 20 28	141	48	0·948	0·941	12 40 02	170	53	115
Canterbury	10 04 41	119	40	315	11 21 31	142	49	0·971	0·970	12 41 58	172	54	113
Cardiff	10 00 21	114	37	316	11 15 35	135	46	0·973	0·972	12 35 08	163	53	118
Carlisle	10 04 39	118	36	307	11 17 48	138	44	0·879	0·854	12 34 31	164	50	122
Carmarthen	09 59 40	113	36	315	11 14 18	133	45	0·965	0·962	12 33 22	160	52	120
Chester	10 02 35	116	36	311	11 16 54	137	45	0·926	0·913	12 35 08	164	51	120
Cork	09 56 01	108	33	316	11 09 04	127	43	0·966	0·964	12 27 04	152	51	124
Coventry	10 03 05	117	38	313	11 18 21	138	47	0·945	0·937	12 37 29	167	52	117
Cowes	10 01 34	116	38	317	11 17 54	137	48	0·991	0·993	12 38 23	167	54	115
Crediton	09 59 10	113	37	318	11 14 37	133	47	0·994	0·995	12 34 37	162	53	118
Derby	10 03 38	117	37	312	11 18 35	139	46	0·931	0·920	12 37 18	167	52	118
Devizes	10 01 28	115	38	316	11 17 12	137	47	0·975	0·975	12 37 06	165	53	117
Doncaster	10 04 43	119	38	310	11 19 22	140	46	0·914	0·898	12 37 36	168	51	118
Douglas	10 02 16	115	35	309	11 15 27	135	44	0·902	0·882	12 32 35	161	50	122
Dover	10 04 48	119	40	315	11 21 49	142	49	0·975	0·975	12 42 25	173	54	113
Dublin	09 59 41	112	34	312	11 12 47	131	43	0·925	0·912	12 30 14	157	50	123
Dumfries	10 04 14	117	35	306	11 17 03	137	44	0·876	0·849	12 33 30	163	49	122
Dundee	10 06 39	119	35	302	11 18 35	139	43	0·838	0·801	12 33 47	164	48	123
Durham	10 05 46	119	37	306	11 19 24	140	45	0·881	0·855	12 36 25	167	50	120
Eastbourne	10 03 18	118	39	317	11 20 11	140	49	0·987	0·989	12 40 56	170	54	113
Edinburgh	10 05 45	118	35	304	11 18 02	138	43	0·852	0·819	12 33 42	164	48	123
Exeter	09 59 14	113	37	318	11 14 46	133	47	0·995	0·997	12 34 50	162	53	118
Exmouth	09 59 15	113	37	319	11 14 53	134	47	0·998	0·999	12 35 03	162	54	117
Fort William	10 05 20	117	34	302	11 16 25	136	42	0·831	0·792	12 30 57	160	47	125
Galway	09 57 16	109	32	313	11 09 25	127	42	0·928	0·916	12 26 14	152	49	126
Glasgow	10 04 45	117	35	304	11 16 48	137	43	0·855	0·823	12 32 22	162	48	124
Gloucester	10 01 45	116	37	315	11 17 06	137	47	0·961	0·957	12 36 33	165	53	117
Great Yarmouth	10 06 43	121	40	311	11 22 52	144	48	0·933	0·922	12 42 16	174	53	114
Hereford	10 01 28	115	37	314	11 16 33	136	46	0·957	0·952	12 35 46	164	52	118
Hexham	10 05 30	119	36	306	11 18 50	139	45	0·876	0·850	12 35 36	166	50	121
Hull	10 05 45	120	38	309	11 20 30	141	46	0·906	0·888	12 38 38	169	51	118
Ilfracombe	09 59 09	113	36	317	11 14 11	133	46	0·982	0·983	12 33 46	161	53	119
Inverness	10 06 57	118	34	300	11 17 47	138	42	0·814	0·770	12 31 49	162	47	125
Ipswich	10 05 32	120	40	313	11 21 54	143	48	0·949	0·943	12 41 42	172	53	114
Kendal	10 04 05	117	36	308	11 17 41	138	45	0·894	0·873	12 34 57	164	50	121
Kirkwall	10 10 16	122	35	296	11 20 06	141	41	0·773	0·719	12 32 42	165	46	126

Times are given in British summer time (BST)

Table 1: Partial eclipse local circumstances

Place	First Contact				Maximum Eclipse					Fourth Contact			
	Time	Az	Alt	PA	Time	Az	Alt	Mag	Obs	Time	Az	Alt	PA
	h m s	°	°	°	h m s	°	°			h m s	°	°	°
Lancaster	10 03 42	117	36	309	11 17 29	138	45	0·902	0·883	12 34 59	164	50	121
Launceston	09 58 19	112	36	319	11 13 33	132	46	0·999	0·999	12 33 30	160	53	119
Leeds	10 04 36	118	37	309	11 18 55	139	46	0·906	0·888	12 36 48	167	51	119
Leicester	10 03 44	118	38	312	11 19 01	139	47	0·938	0·929	12 38 02	168	52	117
Lerwick	10 13 30	125	35	292	11 22 38	145	41	0·741	0·679	12 34 07	168	45	126
Limerick	09 56 50	109	33	314	11 09 26	127	42	0·945	0·937	12 26 50	152	50	125
Lincoln	10 04 57	119	38	310	11 19 59	141	47	0·921	0·907	12 38 34	169	52	117
Liverpool	10 02 49	116	36	311	11 16 59	137	45	0·920	0·905	12 35 02	164	51	120
Llandrindod Wells	10 01 01	114	36	314	11 15 45	135	46	0·952	0·947	12 34 42	163	52	119
London (Central)	10 03 31	118	39	315	11 19 49	140	48	0·968	0·966	12 39 54	169	53	115
Londonderry	10 00 57	112	33	308	11 12 40	131	42	0·881	0·856	12 28 29	156	48	126
Lundy Island	09 58 35	112	36	317	11 13 27	132	46	0·984	0·985	12 32 56	159	53	119
Luton	10 03 38	118	39	314	11 19 37	140	48	0·958	0·953	12 39 21	169	53	115
Manchester	10 03 34	117	37	310	11 17 53	138	46	0·916	0·901	12 35 58	165	51	119
Margate	10 05 09	120	40	314	11 22 01	143	49	0·967	0·965	12 42 24	173	54	113
Merthyr Tydfil	10 00 29	114	37	315	11 15 30	135	46	0·966	0·963	12 34 49	162	53	119
Middlesborough	10 05 51	119	37	307	11 19 43	141	45	0·885	0·862	12 36 58	167	50	120
Minehead	09 59 47	113	37	317	11 15 04	134	46	0·982	0·983	12 34 48	162	53	118
Morecambe	10 03 38	117	36	309	11 17 23	137	45	0·902	0·883	12 34 52	164	50	121
Newbury	10 02 11	116	38	316	11 18 08	138	48	0·972	0·972	12 38 07	167	53	116
Newcastle	10 06 00	119	37	306	11 19 28	140	45	0·875	0·849	12 36 18	167	50	120
Newry, Co. Down	10 00 42	113	34	310	11 13 16	132	43	0·902	0·883	12 29 59	157	49	124
Northampton	10 03 30	118	38	313	11 19 05	139	47	0·949	0·942	12 38 28	168	53	116
Norwich	10 06 16	121	39	311	11 22 16	143	48	0·933	0·923	12 41 36	173	53	115
Nottingham	10 04 01	118	38	311	11 19 03	139	47	0·929	0·917	12 37 46	167	52	118
Okehampton	10 58 46	112	36	318	11 14 06	133	46	0·995	0·996	12 34 04	161	53	118
Omagh	10 00 27	112	33	309	11 12 26	131	42	0·892	0·869	12 28 35	156	49	125
Oxford	10 02 38	117	38	315	11 18 24	138	47	0·962	0·959	12 38 07	167	53	116
Perth	10 06 09	118	35	303	11 18 00	138	43	0·840	0·804	12 33 12	163	48	124
Portree	10 05 18	116	33	301	11 15 39	135	41	0·817	0·774	12 29 26	159	46	127
St. Helier, Jersey	09 59 14	113	38	322	11 16 02	134	49	0·991	0·993	12 37 28	165	55	114
St. Peter Port, Guernsey	09 59 01	113	38	322	11 15 32	134	48	0·998	0·999	12 36 42	164	55	115
Salisbury	10 01 22	115	38	317	11 17 20	137	47	0·983	0·984	12 37 29	166	54	116
Scarborough	10 06 19	120	38	307	11 20 38	142	46	0·892	0·870	12 38 16	169	51	118
Sheffield	10 04 10	118	37	310	11 18 48	139	46	0·918	0·904	12 37 07	167	51	118
Shrewsbury	10 02 10	116	37	313	11 16 50	136	46	0·939	0·929	12 35 31	164	52	119
Skegness	10 05 49	120	39	310	11 21 10	142	47	0·921	0·907	12 39 55	171	52	116
Sligo	09 59 05	110	33	310	11 10 52	129	42	0·901	0·881	12 27 01	153	49	126
Southampton	10 01 36	116	38	317	11 17 48	137	48	0·987	0·988	12 38 09	167	54	115
Stoke-on-Trent	10 03 04	117	37	312	11 17 45	138	46	0·933	0·918	12 36 17	165	52	119
Stornoway	10 06 24	117	33	299	11 16 04	135	40	0·796	0·748	12 29 00	159	46	128
Stranraer	10 02 45	115	35	307	11 15 16	135	43	0·882	0·857	12 31 37	160	49	123
Stratford-on-Avon	10 02 39	117	38	314	11 18 00	138	47	0·951·	0·945	12 37 18	166	52	117
Swindon	10 01 53	116	38	315	11 17 35	137	47	0·969	0·967	12 37 21	166	53	117
Tenby	09 59 06	112	36	316	11 13 41	132	45	0·970	0·968	12 32 47	160	52	120
Tiverton	09 59 27	113	37	318	11 14 54	134	47	0·990	0·992	12 34 51	162	53	118
Tunbridge Wells	10 03 38	118	39	316	11 20 18	140	49	0·977	0·977	12 40 45	170	54	114
Warwick	10 02 52	117	38	313	11 18 12	138	47	0·949	0·942	12 37 26	166	52	117
Weston-super-Mare	10 00 26	114	37	316	11 15 49	135	47	0·977	0·977	12 35 33	163	53	118
Wexford	09 58 15	111	34	315	11 11 51	130	44	0·953	0·947	12 30 02	156	51	122
Weymouth	10 00 11	114	37	318	11 16 10	135	47	0·997	0·999	12 36 33	164	54	116
Workington	10 03 45	117	36	308	11 16 53	137	44	0·887	0·864	12 33 44	163	50	122
Wrexham	10 02 18	116	36	312	11 16 41	136	45	0·930	0·918	12 35 02	164	51	120
Wick	10 09 20	121	35	297	11 19 37	141	42	0·787	0·737	12 32 48	164	46	126
York	10 05 15	119	37	309	11 19 37	140	46	0·902	0·883	12 37 27	168	51	119

Times are given in British summer time (BST)

Magnetic north is not the same as true north. *Polaris*, the Pole Star, lies within 1° of true north whereas magnetic north will lie some 5° west of true north for most parts of the UK by 1999. True north corresponds to an azimuth of 0°, east to 90°, south to 180° and west to 270°. Hence at the start of the eclipse, the Sun lies at an azimuth of 119° which corresponds to approximately east south east. The Sun is at an altitude (Alt) of 39° that is to say that it is 39° above the horizon. The sea-level horizon has an altitude of 0° and the point directly overhead, known as the zenith, has an altitude of 90°. The silhouette of the Moon first appears on the Sun's disk at a position angle (PA) of 313°. In order to understand the term position angle, let us make an analogy between position angle and the direction of the hour hand on a clock face at different times.

Let us suppose that a clock face is superimposed on the Sun's disk with the twelve o'clock position pointing towards the zenith or the point in the sky directly above your head. This point has a position angle of 0°. Moving anticlockwise to the nine o'clock position, the hour hand now points to a position angle of 90°. Continuing the clock face analogy, the hour hand at six o'clock corresponds to 180° and similarly three o'clock corresponds to 270°. Therefore, 313° corresponds to the position of the hour hand at 1^h 34^m. Please refer to the central panel of Figure 3 for a graphical display of position angle.

The next four columns refer to the eclipse circumstances at the moment of maximum eclipse. This occurs at 11^h 20^m 28^s BST when the Sun lies at an azimuth of 141°, roughly south east, at an altitude of 48°. The next two columns are the magnitude (Mag) and the obscuration (Obs) of the eclipse. The magnitude of the eclipse refers to the fraction of the solar diameter covered by the Moon at the moment of greatest eclipse expressed in terms of the solar diameter. This quantity can be greater than one if the apparent diameter of the Moon is greater than that of the Sun. For Cambridge, the magnitude of the eclipse is 0·948. The obscuration is the fraction of the surface of the solar disk covered by the Moon and takes a maximum value of one. It is the obscuration of the Sun which has been plotted in Figure 4. At maximum eclipse 0·941 or 94·1% of the Sun is obscured as seen from Cambridge.

The final four columns refer to fourth contact. Fourth contact or the end of the eclipse occurs at 12^h 40^m 02^s BST when the Sun is at an azimuth of 170°, almost due south at an altitude of 53°. The silhouette of the Moon disappears from the Sun's disk at a position angle of 115°, the position the hour hand of our imaginary clock would occupy at 08^h 10^m. The partial eclipse as seen from Cambridge is over.

10 — Local circumstances of the total eclipse

The path of totality over the Scilly Isles, Cornwall, Devon and the Channel Islands is shown in Figure 5. The first land to see totality is the Scilly Isles. If you are on the UK mainland south of a line joining Port Isaac in Cornwall and Teignmouth in Devon, you will see a total eclipse. The duration of the eclipse will be very short if you are only just south of this line. For example, Teignmouth will see only 14^s of totality. Moving south to Penzance and Falmouth the duration of totality increases to 2^m 06^s. At the southern edge of the track, St. Anne, on the Channel Island of Alderney, will see 1^m 47^s of totality. Guernsey and Jersey are south of the southern limit of totality although Guernsey should see a fine display of Baily's Beads.

Let us suppose we are observing the total eclipse from Falmouth. The information in Table 2 can be explained following the method of Section 9 by tracing the course of the eclipse as viewed from Falmouth. The first column of Table 2 is the location. The next four columns refer to first contact and provide the same information as Table 1. Hence, the start of the eclipse or the beginning of the penumbral phase occurs at 09^h 57^m 06^s BST. The Sun lies at an azimuth of 111° and an altitude of 36°. The silhouette of the Moon first appears at a position angle (PA) of 320° corresponding to the hour hand position at 1^h 20^m on the imaginary clock face introduced in Section 9.

Over the next hour and a quarter more and more of the Sun's disk is covered by the Moon. The sixth column in Table 2 is the time of second contact or the beginning of totality which occurs at 11^h 11^m 15^s BST. The next three columns refer to maximum eclipse. The time of maximum eclipse occurs at 11^h 12^m 18^s BST when the Sun lies in a south-easterly direction at an azimuth of 130° and an altitude of 46°.

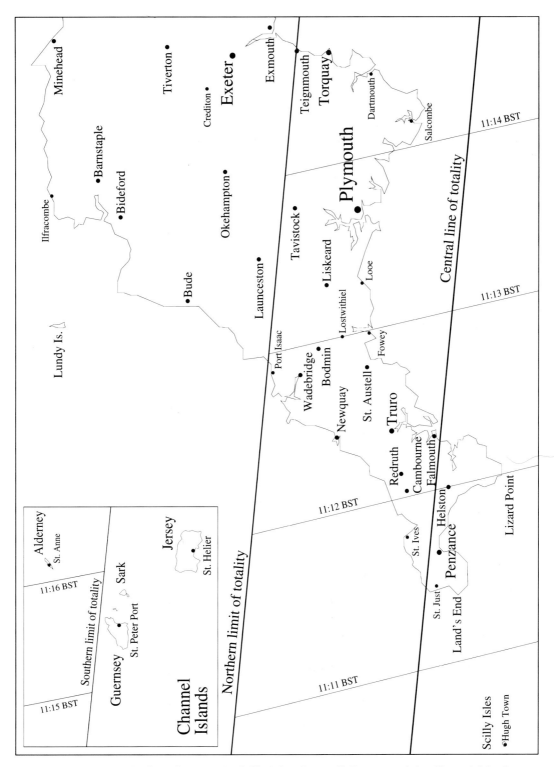

Figure 5: The path of totality over the Scilly Isles, Cornwall, Devon, and the Channel Islands. The northern and southern limits of totality are shown where relevant as well as the central line of the eclipse. The lines crossing the diagram north to south link together locations sharing the same time of maximum eclipse. The time of maximum eclipse is given for each of these lines. A total eclipse will be visible for locations on the UK mainland south of a line between Port Isaac and Teignmouth. The Scilly Islands and Alderney in the Channel Islands will also experience totality.

Coastal outlines © Crown copyright

Table 2: Total eclipse local circumstances

Place	First Contact				Second Contact	Maximum Eclipse				Third Contact	Fourth Contact			
	Time	Az	Alt	PA	Time	Time	Az	Alt	Dur	Time	Time	Az	Alt	PA
	h m s	°	°	°	h m s	h m s	°	°	m s	h m s	h m s	°	°	°
Ashburton	09 58 47	112	37	319	11 13 52	11 14 20	133	47	0 56	11 14 48	12 34 30	161	54	118
Blisland	09 57 53	111	36	319	11 12 30	11 13 03	131	46	1 07	11 13 37	12 33 00	159	53	119
Bodmin	09 57 48	111	36	319	11 12 17	11 12 59	131	46	1 23	11 13 40	12 32 57	159	53	119
Brixham	09 58 55	113	37	319	11 13 49	11 14 37	133	47	1 35	11 15 24	12 34 56	162	54	117
Buckfastleigh	09 58 44	112	37	319	11 13 40	11 14 16	133	47	1 12	11 14 52	12 34 28	161	54	118
Callington	09 58 13	112	36	319	11 12 57	11 13 33	132	46	1 12	11 14 09	12 33 36	160	53	118
Camborne	09 56 58	110	35	320	11 11 00	11 12 02	130	46	2 04	11 13 05	12 32 03	158	53	119
Dartmeet	09 58 42	112	36	319	11 13 53	11 14 11	133	47	0 35	11 14 29	12 34 19	161	54	118
Dartmouth	09 58 47	112	37	319	11 13 36	11 14 28	133	47	1 44	11 15 20	12 34 48	162	54	117
Devonport	09 58 13	112	36	319	11 12 47	11 13 40	132	46	1 46	11 14 33	12 33 51	160	54	118
Falmouth	09 57 06	111	36	320	11 11 15	11 12 18	130	46	2 06	11 13 21	12 32 25	158	54	119
Fowey	09 57 44	111	36	320	11 12 04	11 13 00	131	46	1 53	11 13 57	12 33 06	159	54	119
Gorran Haven	09 57 29	111	36	319	11 11 43	11 12 44	131	46	2 03	11 13 46	12 32 51	159	54	119
Hayle	09 56 48	110	35	320	11 10 47	11 11 50	130	46	2 06	11 12 53	12 31 51	157	53	119
Helston	09 56 52	110	36	320	11 10 57	11 12 00	130	46	2 06	11 13 03	12 32 05	158	54	119
Hugh Town	09 55 41	109	35	321	11 09 34	11 10 28	128	45	1 46	11 11 21	12 30 24	155	53	120
Ivybridge	09 58 28	112	36	319	11 13 11	11 14 00	132	47	1 41	11 14 51	12 34 14	161	54	118
Kingsbridge	09 58 32	112	37	320	11 13 12	11 14 10	133	47	1 56	11 15 08	12 34 30	161	54	117
Kingswear	09 58 48	112	37	319	11 13 38	11 14 30	133	47	1 44	11 15 22	12 34 50	162	54	117
Land's End	09 56 22	110	35	321	11 10 19	11 11 20	129	46	2 03	11 12 21	12 31 21	157	53	120
Liskeard	09 58 02	112	36	319	11 12 36	11 13 19	132	46	1 25	11 14 01	12 33 21	160	53	118
Lizard	09 56 48	110	36	321	11 11 02	11 12 01	130	46	1 59	11 13 00	12 32 13	158	54	119
Looe	09 57 54	111	36	320	11 12 19	11 13 16	131	46	1 52	11 14 12	12 33 24	160	54	118
Lostwithiel	09 57 46	111	36	319	11 12 10	11 13 00	131	46	1 41	11 13 51	12 33 03	159	53	119
Mevagissey	09 57 31	111	36	320	11 11 45	11 12 46	131	46	2 00	11 13 46	12 32 52	159	54	119
Mullion	09 56 49	110	36	321	11 10 58	11 12 00	130	46	2 03	11 13 01	12 32 09	158	54	119
Newquay	09 57 23	111	36	320	11 11 35	11 12 26	131	46	1 42	11 13 17	12 32 22	158	53	119
Newton Abbot	09 58 57	113	37	319	11 14 13	11 14 33	133	47	0 40	11 14 53	12 34 45	162	54	118
Newton Ferrers	09 58 18	112	36	320	11 12 53	11 13 49	132	47	1 53	11 14 46	12 34 04	160	54	118
Padstow	09 57 39	111	36	319	11 12 09	11 12 43	131	46	1 06	11 13 16	12 32 35	159	53	119
Penzance	09 56 37	110	35	320	11 10 35	11 11 37	130	46	2 06	11 12 40	12 31 38	157	53	119
Perrenporth	09 57 14	111	36	320	11 11 21	11 12 18	130	46	1 54	11 13 15	12 32 16	158	53	119
Plymouth	09 58 14	112	36	319	11 12 50	11 13 41	132	46	1 42	11 14 32	12 33 51	160	54	118
Polperro	09 57 51	111	36	320	11 12 14	11 13 10	131	46	1 53	11 14 07	12 33 18	159	54	118
Port Isaac	09 57 49	111	36	319	11 12 35	11 12 54	131	46	0 39	11 13 13	12 32 47	159	53	119
Portreath	09 57 01	110	35	320	11 11 03	11 12 04	130	46	2 02	11 13 05	12 32 04	158	53	119
Prawle Point	09 58 29	112	37	320	11 13 09	11 14 11	133	47	2 04	11 15 13	12 34 35	161	54	117
Redruth	09 57 01	110	36	320	11 11 05	11 12 07	130	46	2 04	11 13 09	12 32 09	158	53	119
Roche	09 57 37	111	36	320	11 11 56	11 12 47	131	46	1 42	11 13 38	12 32 47	159	53	119
St. Agnes	09 57 09	111	36	320	11 11 14	11 12 13	130	46	1 57	11 13 12	12 32 11	158	53	119
St. Anne, Alderney	09 59 37	114	38	321	11 15 15	11 16 08	135	48	1 47	11 17 02	12 37 12	164	55	115
St. Austell	09 57 41	111	36	319	11 12 03	11 12 52	131	46	1 38	11 13 41	12 32 52	159	53	119
St. Columb Major	09 57 32	111	36	319	11 11 50	11 12 39	131	46	1 38	11 13 28	12 32 36	159	53	119
St. Ives	09 56 46	110	35	320	11 10 43	11 11 46	130	46	2 05	11 12 48	12 31 44	157	53	119
St. Just	09 56 30	110	35	320	11 10 24	11 11 27	129	46	2 05	11 12 30	12 31 26	157	53	120
St. Stephen	09 57 29	111	36	320	11 11 44	11 12 40	131	46	1 53	11 13 36	12 32 41	159	53	119
Salcombe	09 58 29	112	37	320	11 13 08	11 14 09	133	47	2 01	11 15 09	12 34 31	161	54	117
Tavistock	09 58 27	112	36	319	11 13 27	11 13 49	132	46	0 44	11 14 11	12 33 53	160	53	118
Teignmouth	09 59 05	113	37	319	11 14 35	11 14 42	133	47	0 14	11 14 49	12 34 54	162	54	117
Torquay	09 58 59	113	37	319	11 14 03	11 14 39	133	47	1 12	11 15 15	12 34 55	162	54	117
Totnes	09 58 45	112	37	319	11 13 37	11 14 22	133	47	1 31	11 15 07	12 34 38	161	54	118
Trenance	09 57 29	111	36	319	11 11 48	11 12 33	131	46	1 30	11 13 18	12 32 27	158	53	119
Truro	09 57 15	111	36	320	11 11 23	11 12 24	130	46	2 01	11 13 24	12 32 26	158	53	119
Wadebridge	09 57 43	111	36	319	11 12 11	11 12 50	131	46	1 19	11 13 29	12 32 46	159	53	119
Yelverton	09 58 26	112	36	319	11 13 15	11 13 51	132	46	1 12	11 14 27	12 33 58	160	54	118

Times are given in British summer time (BST)

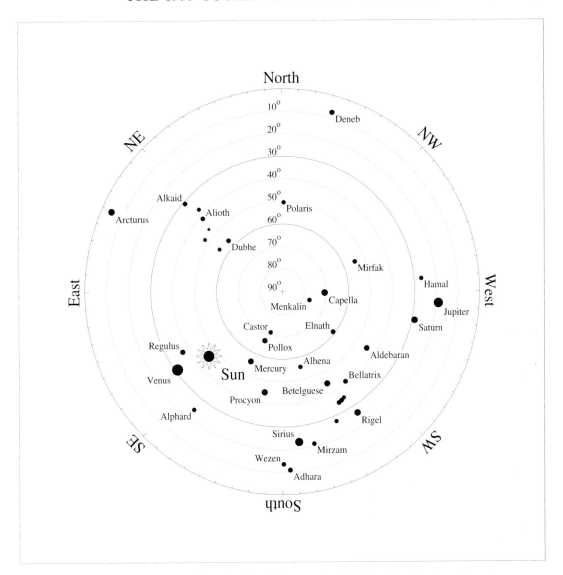

Figure 6: An overhead view of the sky as it will appear from Falmouth at the time of maximum eclipse. The centre of the plot represents the zenith and altitudes are marked in intervals of 10°. Around the edge of the diagram is the horizon where the main compass points are given with tick marks every 5° in azimuth. This diagram is applicable to all those parts of the UK in the path of totality. All the planets visible to the naked eye and the main stars in the constellations of Ursa Major and Orion are plotted in addition to those stars brighter than magnitude 2·0. The size of the symbol is related to the apparent magnitude of the object, the larger the filled circle the brighter the object.

The next column is the duration of totality which is 2^m 06^s for Falmouth. This quantity has been calculated on the assumption that the Moon is a smooth sphere. However, the topography of the lunar limb, in particular the presence of deep valleys, can reduce the obscuring effect of the Moon and hence decrease the duration of totality by several seconds.

As the sky darkens, some of the planets will become visible. Mars is below the horizon at the time of totality and Jupiter and Saturn, at magnitudes −2·5 and −0·1 respectively, will be setting in the west and may be difficult to identify. However, Mercury and Venus should be seen quite easily. Mercury is 18° west of the Sun with a magnitude of +0·6 whereas Venus is 15° east of the Sun with a magnitude of −4·1. For many people, this will be their first view of Mercury, a planet which never strays far from the Sun as seen in our skies.

Similarly, some bright stars will also become visible. At the time of the eclipse, the Sun is in the constellation of Cancer, the Crab, a constellation devoid of bright stars. However, bright stars which are relatively close to the eclipsed Sun include *Sirius, Regulus, Castor, Pollux* and *Procyon.*

A simplified sky chart is given in Figure 6. The centre of the plot represents the zenith and altitudes are marked in intervals of 10°. Around the edge of the diagram is the horizon where the main compass points are given with tick marks every 5° in azimuth. Stars brighter than magnitude 2·0 are plotted as well as those planets visible to the naked eye. To identify stars and planets during the total eclipse, hold the map above your head with North pointing in the direction of true north. If you are facing south, the Sun should be on your left hand side in front of you. Although this diagram has been produced for the skies above Falmouth at the time of maximum eclipse, it should be applicable to those parts of the UK on the path of totality. The visibility of some or all of these objects will depend on local weather conditions including the presence of haze or mist.

Returning to Table 2, the next column is the time of third contact or the end of totality which occurs at 11^h 13^m 21^s BST. The last four columns refer to fourth contact or the end of the penumbral phase of the eclipse. This occurs at 12^h 32^m 25^s BST. The Sun is now at an azimuth of 158°, a south south easterly direction and an altitude of 54°. The silhouette of the Moon leaves the Sun's disk at a position angle of 119° or the position of the hour hand at 08^h 02^m on our imaginary clock face. The eclipse for observers in Falmouth is now complete.

In the Channel Islands, the eclipse is total for those observers on Alderney, the most northerly member of these islands. Table 2 shows that the eclipse begins at 09^h 59^m 37^s BST. The Sun lies at an azimuth of 114° and an altitude of 38°. The silhouette of the Moon appears at a position angle of 321°. Second contact occurs at 11^h 15^m 15^s BST and totality lasts for 1^m 47^s. The duration of totality is somewhat shorter than for the mainland as Alderney lies further away from the central line of the eclipse. By maximum eclipse at 11^h 16^m 08^s BST, the Sun has reached an azimuth of 135° and an altitude of 48°. Third contact occurs at 11^h 17^m 07^s BST and the eclipse ends at 12^h 37^m 12^s BST when the Sun lies at an azimuth of 164° and an altitude of 55°. The silhouette of the Moon leaves the Sun's disk at a position angle of 115°.

11 — Observing the eclipse

There are many ways to observe a total eclipse of the Sun. The primary consideration should always be safety. Jeopardizing your eyesight to watch the eclipse using inadequate precautions can never be justified by the rarity of such an event over the UK. **The only time it is safe to look directly at the Sun is during totality. At any other time, the ultraviolet and infrared radiation will damage your eyesight even though you may not feel any discomfort. Do not stare at the Sun.** Observing the Sun with any form of optical aid is potentially very dangerous unless you know what you are doing. If you are not sure about what you are attempting to do, the best advice is not to do it at all!

There are several methods of observing the partial phases of the eclipse safely. The simplest method is pinhole projection which requires two pieces of stiff cardboard. A clean pinhole is punched in one piece of card. Standing with your back to the Sun, arrange the pieces of card so that sunlight passes through the hole in the first piece of card and is projected on to the second piece of card held about a metre from the first. An inverted image of the Sun will be seen. Make sure the pinhole is not too wide otherwise no image will be formed. The size of the image can be adjusted by changing the separation between the two pieces of card. Do **not** look through the pinhole at the Sun. Projecting the Sun using binoculars can also give several people the opportunity to observe the eclipse simultaneously. Keeping the objective cover on one side of the binoculars, a sheet of cardboard is placed around the other objective lens of the binoculars. This acts as a shade for a second sheet of cardboard placed approximately 30 centimetres behind the eyepieces of the binoculars upon which the image of the Sun can be projected. Aligning the binoculars should be done by trial and error and not by looking through them. The image should be focused using the focusing knob and adjusting the distance between the binoculars and the projection screen. A similar method can be adopted for use with a small refracting telescope.

Solar filters made of aluminized mylar can be used. However, they **must** be checked for flaws which might allow direct sunlight to reach your eyes. The best type of filter is one coated with aluminium on both sides of the mylar film. This minimizes the possibility of an alignment of pinhole flaws in the aluminium coatings on each side of the film. Welder's goggles are also suitable for observing the eclipse as long has they have a rating of 14 or higher and have been checked for their infrared transmission characteristics (see Amendment 3 to BS679:1959; "Filters for use during welding and similar industrial operations"). Specialist camera and telescope filters are also available but these are in general expensive and only available from specialist suppliers (see Section 15).

Sunglasses of any type **must not** be used for looking at the Sun. They do not block those wavelengths of light likely to damage your eyes nor do they provide the necessary reduction in the intensity of incoming light. The damaging radiation from the Sun is also unaffected by polaroid sunglasses. Smoked glass can provide reasonable protection if the glass is large enough and the density of the carbon deposit is sufficiently high. Unfortunately, making a uniformly dark filter is difficult and the degree of protection cannot be guaranteed. As a result, this method cannot be recommended. Fully-exposed and developed film should not generally be used. Colour film and chromogenic black and white films are totally unsuitable as harmful infrared radiation is not blocked by the coloured dyes used in these preparations. Under **no** circumstances should gelatin neutral density filters be used. Standard 35 millimetre negatives of any type are physically too small to be used as solar filters. However, certain types of fully exposed and developed black and white film using metallic silver can be used as solar filters. Two layers of film must be used for brief views of the Sun amounting to less than about 30 seconds in duration. If you are unsure about the type of film you have **do not** use it!

For people who wish to retain some record of the eclipse, videography and photography are possible. Like the eye, the detector in the average video camera will be destroyed by exposing it directly to full sunlight. In general, exposure times and aperture settings are dealt with automatically by the electronic metering system within the video camera. However, some form of mylar or glass solar filter capable of cutting the amount of light and heat down by a factor of approximately 100 000, corresponding to a neutral density of 5·0, is required during the partial phases to allow suitable exposures times and aperture stops to be set by the automatic metering system in the camera. During totality, the filter should be removed. Caution must be exercised when aligning the video camera with the Sun.

Many people will try to photograph the eclipse. A normal camera lens with a 50 millimetre focal length will produce an image of the Sun some 0·5 millimetres in diameter. Consequently, a telephoto lens is necessary to give an acceptable image size. A 500 millimetre lens would produce an image of nearly 5 millimetres across. To capture the Sun's corona on film, a lens of approximately 1000 millimetres represents the best compromise between the necessary field of view and an acceptable image size. Photography during partial phases will require some form of filtering. As in the case of videography, a filter which cuts the amount of light and heat down by a factor of 100 000 or so is recommended. Care must be exercised when setting up the camera because using the optics in the camera viewfinder, as you would for normal photography, can damage your eyesight. A similar warning applies to the use of a single lens reflex (SLR) camera as you are using the camera lens to view the scene you are about to photograph. Exposure times will depend on the amount of filtering. Since there is plenty of light, a film speed of ASA 50 to 100 is suitable. Experimenting on an uneclipsed Sun is a good way to gauge exposure times and using a filter of neutral density 5·0 with ASA 100 film at f/8 would probably require an exposure of 1/125 second.

Photography during totality does not require a filter. Assuming you are using ASA 100 film at f/8, prominences could be photographed with an exposure of about 1/60 second. To capture the corona would require an exposure of between 1/8 second and 1/2 second. The key to successful photography is to bracket your exposure by several f-stops. No single exposure time is the correct one. The Sun is not the only object you might photograph during the eclipse. Many photographers try to capture the effect the eclipse has on their surroundings. Here are two pieces of advice from veteran eclipse watchers. If this is your first total eclipse, watch it rather than attempting to photograph it yourself. Professional photographs will be available after the eclipse of a quality exceeding anything most amateurs might achieve. Secondly, for those people who own cameras with a built-in automatic flash gun, don't use it! You will only cause consternation amongst your fellow eclipse watchers when their dark adaption is ruined by the discharge of a flash gun caused by the low light levels during totality. However, if you are seriously considering photographing the eclipse, please consult the books listed in Section 16.

It has been calculated that on average a total eclipse would be visible at a given location on the Earth every 400 years or so. It is therefore an opportunity not to be missed. Wherever you go to watch the eclipse, enjoy this once in a lifetime event but above all else take care.

12 — Weather in the path of totality

The following information is by no means a definitive description of weather conditions over the path of totality, it is simply a guide to likely weather patterns. To see the eclipse properly good weather conditions, particularly minimal cloud cover, are vital. Unfortunately, even in August, the weather in the United Kingdom cannot be predicted with any great accuracy.

The average rainfall in south-west England during August is approximately 100 millimetres. This decreases to around 75 millimetres in the Scilly Isles and around 60 millimetres in the Channel Islands. If rainfall does occur it is likely to be of a showery nature. On average, thunderstorms are likely to occur on two days of the month in the Channel Islands, on one day of the month in Cornwall and on only half a day in the Scilly Isles. May is the sunniest month in the Scilly Isles whereas June is the sunniest month in the Channel Islands and Cornwall. The Channel Islands have the highest average total number of hours of sunshine in August at around 220 hours per month. Similarly, the Scilly Isles have about 205 hours and Devon and Cornwall have about 180 hours. Examining the figures more closely, the Channel Islands have more than 9 hours of sunshine per day for nearly half of August and two thirds of August have more than six hours of sunshine. In Cornwall, only one quarter of the month has more than 9 hours of sunshine a day and half the month has more than 6 hours of sunshine per day.

Much of the sunshine in south-west England and the Channel Islands occurs in the afternoon period for different reasons. Weather records show that during the morning south-west England is affected by cloud around dawn which gradually disperses throughout the day. Much of this area spends 50% of its daylight hours under skies with more than 75% cloud cover. Observers in Cornwall and the Scilly Isles may have their view of the eclipse disrupted by clouds because as much as two thirds of the sky may be affected by cloud at the time of the eclipse. However, the Channel Islands do appear to fare a little better. Records show that the problem here may be early morning fog. Approximately one quarter of the month is affected by such fog. Once this early morning fog has dispersed, the skies are somewhat clearer than the mainland.

On balance, there is probably little to choose between the weather in south-west England and the Channel Islands. Local weather conditions may vary greatly between locations separated by as little as 30 kilometres. As far as Cornwall is concerned, the prospects of seeing the eclipse may improve marginally the further south west you go. In the Channel Islands, Alderney being a relatively small, flat island may fare quite well.

13 — Future eclipses of the Sun in the UK

Although the Faroe Islands will see a total eclipse of the Sun on the morning of 20th March 2015 and the Channel Islands will see another total eclipse of the Sun early on the morning of 3rd September 2081, the eclipse of 11th August 1999 will be the last opportunity to see a total eclipse of the Sun in the UK mainland for another 90 years. On 23rd September 2090, the south-western tip of the Irish Republic, south-west England, most of the south coast of England and the Channel Islands will again see a total eclipse of the Sun just before sunset. As the path of totality is more than 240 kilometres wide, totality will be seen by much of northern France as well. Most of the UK will also witness a partially-eclipsed Sun at sunset.

The next two total eclipses will occur in quick succession, the first on the morning of the 3rd June 2133 over the Outer Hebrides, the Shetland Islands and the north-west tip of Scotland and the second shortly after sunrise on 7th October 2135 over central and southern Scotland and north-east England. Seven years later, the Channel Islands will see yet another total eclipse of the Sun early on the morning of 14th June 2142. After a further nine years, another total eclipse will occur in the early evening of 14th June 2151 which will be seen as total from Northern Ireland, north Wales, south-west Scotland, northern England, the Midlands and East Anglia.

On a shorter timescale, there are three solar eclipses worthy of some discussion in the next 30 years or so. The solar eclipse of 31st May 2003 is an annular eclipse. Parts of northern Scotland will see it as an annular eclipse but for most observers on the UK mainland the eclipse will be seen in a partial phase during sunrise. The skies over the UK and the Irish Republic will darken quite substantially on the morning of 20th March 2015. The track of this total eclipse passes over the northern Atlantic Ocean between Iceland and northern Scotland reaching as far as the North Pole. The obscuration of the Sun will be about 85% as seen from south-east England rising to around 90% for north-west Scotland. Another total eclipse track passing to the west of the Irish Republic is that of 12th August 2026 which will again bring a substantial darkening of the sky just before sunset. Details of these and other eclipses will be available in future publications of H.M. Nautical Almanac Office.

14 — A quick summary

If you want to use this booklet as a quick reference to discover what you are likely to see at a specific location in the UK, the Irish Republic or the Channel Islands, please follow these steps:

- Identify your location on Figure 4. By comparing your geographical position with the nearly horizontal lines on the map, you can determine how much of the Sun will be obscured. You will also be able to estimate the time of maximum eclipse from the near vertical lines.

- If you think the location from which you will observe the eclipse lies close to path of totality, confirmation can be obtained using Figure 5. If you are in the Scilly Isles, on Alderney or in south-west England south of the northern limit of totality you will see a total eclipse of the Sun.

- If you are likely to see a partial eclipse, look in Table 1 for a city or town close to your location. Reading across the table will give you the times and positions of the Sun at the start of the eclipse, maximum eclipse and the end of the eclipse along with the relevant obscuration. The position angle of the appearance and disappearance of the silhouette of the Moon on the Sun's disk is also given.

- If you are likely to see a total eclipse, look in Table 2 for a town or village close to your location. Reading across the table will give you the times and positions of the Sun at the start of the eclipse, the beginning and end of totality and the end of the eclipse. The position angle of the appearance and disappearance of the silhouette of the Moon on the Sun's disk is also given. Figure 6 will give you some idea of what objects will be visible in the sky during totality.

- If you want to see how the eclipse progresses and to visualize the position angle information provided in Tables 1 and 2, please refer to Figure 3. This diagram may also prove helpful in predicting the appearance of Baily's Beads. They may be visible where there is a sizeable gap between the solid outline of the Moon and the photospheric disk of the Sun.

To observe the eclipse safely, **do not** use the following methods to watch the partial phases of the eclipse as eye damage may result:

- Sunglasses of any type
- Gelatin filters
- Fully exposed and developed colour film
- Fully exposed and developed black and white negatives in general
- Smoked glass

The following methods can be used to observe the partial phase but please remember to exercise caution in their use:

- Pinhole projection
- Projection using binoculars or a small telescope

- Welder's goggles rated at 14 or higher
- Aluminized mylar filters

Following these simple rules should minimize the possibility of damaging your eyesight. Even if you are using the right eye protection, do not stare at the Sun for long periods.

15 — The aluminized mylar eclipse viewer

Inside the back cover of this booklet, an eclipse viewer has been provided to allow observation of the partial phase of the eclipse. The viewer is constructed from a sandwich of two layers of aluminized mylar cemented together with the aluminium on the inside. This reduces the intensity of sunlight to a safe level. These eclipse viewers and other solar filters can be obtained from Eclipse99 Limited, Belle Etoile, Rue du Hamel, Castel, Guernsey GY5 7QJ.

Certification under the provisions of the Personal Protective Equipment Regulations has been sought. Viewers provided with this guide will carry the CE mark indicating that this certification has been granted. However, it is important to remember that the mylar can be damaged by puncturing with a sharp object or by heat sufficient to melt the mylar. **Please check the viewer for damage before use**.

16 — Useful references

- *The Cambridge Eclipse Photography Guide* by Jay M. Pasachoff and Michael Covington and published by Cambridge University Press.

- *Totality: Eclipses of the Sun* by Mark Littman and Ken Willcox and published by the University of Hawaii Press.

- *Total Eclipses of the Sun* by J.B. Zirker and published by Van Nostrand Reinhold Limited.

- *UK Solar Eclipses from Year 1 (an anthology of 3,000 years of solar eclipses)* by Sheridan Williams and published by Clock Tower Press.

Printed in the United Kingdom for HMSO
Dd302839   C15   9/96   (32719)